CHANGING T

CHANGING THE SUBJECT

Carole Satyamurti

Oxford   New York

OXFORD UNIVERSITY PRESS

1990

Oxford University Press, Walton Street, Oxford OX2 6DP

Oxford New York Toronto
Delhi Bombay Calcutta Madras Karachi
Petaling Jaya Singapore Hong Kong Tokyo
Nairobi Dar es Salaam Cape Town
Melbourne Auckland

and associated companies in
Berlin Ibadan

Oxford is a trade mark of Oxford University Press

First published in Oxford Poets
as an Oxford University Press paperback 1990

British Library Cataloguing in Publication Data
Satyamurti, Carole
Changing the subject.—(Oxford poets)
I. Title
821.914
ISBN 0–19–282738–3

Library of Congress Cataloging in Publication Data
Data available

Typeset by Wyvern Typesetting Ltd
Printed in Great Britain by
J. W. Arrowsmith Ltd., Bristol

For Sathya

ACKNOWLEDGEMENTS

Thanks are due to the editors of the following publications in which some of these poems first appeared: *Poetry Review*, *Skoob Review*, *Encounter*, *Illuminations*, *Ambit*, *London Magazine*, *Verse*, *The North*, *Oxford Poetry*, *Prospice*, *The Rialto*, *The Observer*, *Vision On*, *Poetry Book Society Anthology 1987–8*, ed. Gillian Clarke (Hutchinson, 1987), *First and Always*, ed. Lawrence Sail (Faber and Faber, 1988), *Arvon Foundation Poetry Competition 1987 Anthology*, eds. Ted Hughes and Seamus Heaney.

CONTENTS

DRIVING THROUGH FRANCE

Between croissants and croque monsieur,
in the time it takes Madame Du Plessis
to wash her coffee bowl,
take up her basket
and walk down to the shops and back,
greeting her neighbours occasionally

we have covered 174 kilometres,
passed through 23 villages
in which 237 women, 84 men and 30 dogs
were walking to the shops, or back
—and have not moved,
nor greeted anyone.

*

When I was about eight, I thought
what luck that I was born
English—not foreign
like most people in the world.

Now, flashing through yet another
undistinguished village, it strikes me
that, for some, the centre of the world
is this strip of houses called Rièstard;

whereas I know it is London
or, rather, Crouch End
or, currently,
this Ford Fiesta.

*

Here are three images:
a round bed of sunflowers in a wheat-field;
an albino boy leaning on a wall;
a pair of gates shaped like swans embracing.

Perhaps it wasn't really a flower-bed,
nor the boy really albino, nor the gates
the shape of swans. Perhaps
speed made them remarkable.

I can't return them.
I could embroider them to arbitrary life;
or file them, tokens that something happened,
like the programme from the son et lumière.

LASCAUX

Sensing her dwindling weight,
she gropes forward, looking for the place
where she can leave her mark
—this cave. Its shapes excite her,
she dreams me. Exquisite generosity of line
licks her vision round cool curves of rock.

Her fierce breath fires me.
I dance, I grow full-bellied,
I am all horses.
Her people come, see me as I was meant.
I feel worshipped.

In time
the shallow hand-held lamps go cold.
I gallop through millennia
of dark and silence.
Then—the boys, the priest,
heat, light, crowds ...

They love me differently;
they are noisy with possibilities.
I am: a present from the Ice Age,
        icon,
        matrix,
        source of income,
        range of pigments,
        stuff of reputation,
        hypotheses to fuel their quarrelling.
I am too old to make them see.

I grow hot, weak, my colours fade,
green sickness blurs my limbs.
This is something they can understand;
they turn the spot-lights out.

Time stretches, sinks to its old pulse.
I hear the tinkle of remote debates.
Each time they think of me
I change a little.

ON NOT BEING A NATURE POET

Picking up a small, white feather
I note its symmetry, each tiny rib
knowing its proper measure.

I hold it in my palm, and speculate
how many I would have to balance there
before I'd feel the weight.

I see its consummate design, spare
curve like a careful hand, repelling water,
nurturing warm air.

Stroking along its spine, I like to sense
the finger-numbing softness near the root
change to resilience.

But it doesn't move me; I can't say
I love it. As I've written this, the wind
has carried it away.

STRAWBERRIES

I'm spun through time widdershins
to a room lumbered
with a childhood's furniture:
stout mahogany, teak that ousted it,
boxy armchairs, brocatelle
that smartened them as my parents
more or less kept pace with progress
—all there, sharing head-space,
colours mixed by memory to a common brown,

though outside, through French windows,
stand the well-mannered, dusty greens
of a town garden—where I hear
heels clack along the path: my mother
back from a hundred shopping trips
with some treat tucked into her basket;
and where I see my father, the day
he ran to buy me strawberries
and found it was a rag-and-bone man.

As she comes, my vague unease dissolves
—home will be home again;
and, as he does, the wrench
of wishing I could reach into his pocket,
show him the treat he didn't know he'd brought.

SHOWING

We brought our mothers' photos in
and had a show. We propped them
in a row along a shelf,
scrutinized their conformation:

Christine's, who went out to work
and voted Labour,
a straight-backed Scottish terrier,
tough and guarded.

Mrs Ascoli's borzoi profile
—taut nerves and tragedy;
exquisite in pearls and flowered straw,
head angled in the subtlest condescension.

Mary's, old and sad
—a bloodhound, hair in loops.
Jane's stocky, cheerful pug-dog of a mother
four-square with a golf-club.

Only mine was human
—a musical-box dancer
radiant in a thousand sequins.
They all agreed she was the prettiest.

Then I was ashamed I'd brought that one
—she and my father at the Ladies' Night,
eyes shining at each other;
the one that looked like history.

BIRTH RITE

Since I've not known another birth
this surgery seems natural.
I've left my home
and have come here
to be prepared.

You are my grail
and I must purify myself
—be stripped, shaved, emptied,
wrapped in white—
before I gain you.

Soon you'll be lifted
from the domain of wishes
and we who have been so intimate
will touch at last. Perhaps
we'll be awkward with each other.

Hiss of trolley wheels,
haze of lights . . . I'm drawn
through deepest passages,
protected, raised; someone
holds my hand perfectly.

To be reborn with you
I shed responsibility,
my social face,
speech, consciousness.
I reach back to the dark.

PULLING AWAY

Your first journey alone.
I've seen you to your seat,
stowed your bag, fussed
over coat and book and sandwiches.
You told me not to ask another passenger
to help you out at York;
you'll manage (but the bag's too heavy),
someone will help, you say.

Of course they will.
Your at-homeness in the world
is something I took thirty years to learn.
You assume goodwill, are scrupulous
in not abusing it. In you
grace, joy, directness
—all the rich subtleties of spirit—
resolve into clear light.

Growing up may be commonplace,
but sometimes you make me catch my breath,
you are so right—my miracle
that started when I held you, new,
and flawed, so that I dreamed for days
of losing you and making a fresh start.
Each uncomplaining step since then
has cost you many times the going rate.

I pace the platform.
As though this were a kind of summing up,
the window frames dozens of images
—you tease a kitten, build a blanket house,
blow candles, cry about hair cut too short . . .
and you now, grave, too small for your seat.
We mouth silly messages. Surely it's time.
You're a patience expert.

You've waited too long for this journey.
But now, the whistle, final slams of doors.
You wave, excited now, blowing kisses
that hover in the space between us.
Am I the child, or you—need, being needed
inextricable? Your fragility is my excuse.
I am the one afraid to move away,
to let go of your hand.

MOTHER'S DAY

for Phyllida Peake

'You don't care! They always die!
You're never any good with animals.'
Slammed doors,
a storm of tears
gusting up the road to school.

In the still kitchen,
fine-boned Topsy
back from her caesarian,
lies with her surviving kitten,
both too drugged to feed.

It fits my hand,
palpitating, so light.
I stare out at the sooty garden,
cemetery for gerbils, budgies, rabbits,
all my miscarriages of love.

Too much to do.
The spiteful ticking of the clock
reminds me—each day
a fight against chaos,
barely won.

But now I'm on my knees.
My fingers try to be a mother's tongue,
roughing it awake.
It stirs. I clamp
its mouth around her teat.

Its head lolls slack.
I'm hectoring them now,
almost angry at their
easiness with death.
All day, I will them, all day

I'm forcing them together.
The house suspends its breath.
My breasts tingle with desire
for this connection.
And then, it's finished

—mother and baby locked,
Topsy licking, purring.
Here, on my lap,
to weigh against a dozen graves,
this tiny Lazarus.

LADY WILDE TAKES LEAVE
OF HER CRITICS*

'Lasciate ogni speranza voi ch'entrate'

Now it has come to this—it's easy
for you to shun me, tittering
behind your hands—yet you know
I am descended from Dante Alighieri.

You've seen me when I was
magnificent—crowned with laurel,
attended by my sons. All Dublin,
London, sought my hospitality

—my salon was the birthplace
of ideas. We all had genius.
The world that mattered—Browning,
Ruskin, Yeats . . . they all came.

There were no petty differences
of sex or race. Contemptuous
of fashion, I was my own creation,
cheating time with artificial dusk.

I don't give house-room to regret.
I chose the pseudonym 'Speranza'
to live my life by, never looked
for squalor. Now, again, it drags at me:

this great misfortune. You may keep
your purblind pity; my grief
is not for your narrow reasons
—I was always above respectability—

it is because he's so diminished.
I never thought . . . he cannot seem
to turn disgrace to his own ends,
laugh at it, knowing he is immortal.

I try to breathe courage into him
but I've drawn on hope too heavily.
He is an oyster, shell torn away.
I am an empty carapace.

*Lady Wilde was the mother of Oscar Wilde.

WOMAN BATHING IN A STREAM:
Rembrandt

Just 'woman'.
We know it was your Hendrickje,
who bore your daughter,
reared your son,
fed you, clothed and sheltered you,
sat, stood, lay down for you,
and who, even in death,
kept you from creditors.
Almost everything we know of her
is what she did for you.

I'm angry for her
—that you took everything,
made her a vehicle for light,
shadow and reflection
and gave her only anonymity
—as now, in fashion photographs:
dress by Cardin,
hat by David Shilling,
ear-rings, necklace by Adrian Mann
and a model with no name.

Yet I can see how you refused to prettify
the ungainly shift, hoisted to hip level,
thick thighs, peasant forearms, shoulders;
how you seem to have felt their balance,
understood her spirit weight
—painted almost in her idiom.
She must have known—no wonder, then,
the serene half-smile, lack of artifice.
Being so recognized
perhaps made simple fame irrelevant.

THE BALCONY: after Manet

We form a perfect composition,
a triangle, he at the apex;
soft, glutton's hands
smelling of sandalwood and Havanas.
Though I gaze down at the street,
I know how his thumb and index finger
stroke each other, round and round,
oh, so slowly.
A woman's skin: a sheaf of banknotes.

I dig my fingers hard against my fan
to block the screaming.
I could gather up my skirts
and vault the rail;
or leap at him, plunge my nails
into those too easy going eyes.
But I sit here,
tame as this agapanthus in a pot,
central, yet marginal.

My little sister with the holy look
falters on the threshold. Will she
step on to the balcony beside me,
her cachou breath warm on my cheek?
Or will she stay, give him
that second's sweet complicity
for which he waits,
a faint flush rising,
stroking, stroking?

NO MAN'S LAND: PORTRAIT OF HILDA CARLINE*

A difficult woman
—slept, ate at the wrong times,
was deaf to crying babies,
awkward when elegance was wanted.

She withheld applause; flaunted
her own too angular opinions.
Here, she is treated harshly,
uglier than she was.

No mother in this nature, no repose,
no contours lavishly displayed,
curve of lush flank,
promise in an eye's glint;

rather, obstinacy, flint,
cheeks planed by rock-falls;
uncompromised by flattery,
even the hair hard-furrowed,

eyes obsidian, shadowed
in outcrop of brow, nose chiselled
straighter than life
in rage-red symmetry.

Her jaw describes the geometry
of will—as though
it could only crack, spew
the grief of decades into chill air

or, cleansed of flesh, made bare,
all passionate resistance over,
winds might sing through orifice of bone
their wild, impersonal plain-chant.

*Hilda Carline was the first wife of Stanley Spencer. A portrait of her by her brother, Richard Carline, hangs in the Tate Gallery, London.

TWINS

M looks in the long mirror
on her mother's wardrobe door.
'Me get in too!', she says, running,
bumping her nose on glass.

*

They throw puddles at each other.
T has a plastic cup;
M has her two pink hands
scooping, throwing, in one jerky movement.
As each stoops, she sees
her agitated reflection; it is
her sister's face she destroys
as she smashes the water's skin.

*

Men fall for both of them,
or either. They marry two
who've learned the sharp distinction.

*

Three times a week at least
they ring each other,
often at the same moment.
They go through the same menopause,
the same lapses of memory.

*

Synthetic organ music
ushers M towards the furnace.
The curtains open.
T is sliding, falling.
The stroke leaves her left side weak,
her face oddly altered.

REFLECTIONS

Looking for myself,
I creep from one reflection to the next.
I stare; I see
suggestions of my son, my granddaughter.
I'm not there,

though if I should bend this way, and this
couldn't I curve back to the place
where the first mirror surely held me
in perfect, loving, infinite regard?

I'm drawn to any gleaming surface
—the polished floor, a silver horn,
windows in a revolving door.
They're never right, never
that milk-blue light I'm longing for.

Often I'm only smudges,
or scattered by cracks;
but I'm there at least,
I've some hold on the ground inhabited
before I found out what I lacked,
and what the mirror did.
And what the mirror
did.

DISABLED SWIMMERS' NIGHT

I would get given that one.
Bean bag body,
dribbling,
arms jerking.
I know she can't help it
but it gives me the creeps.
Help her in a minute,
just a bit longer,
chatting.

        Any minute, I'll be fish,
        dizzy, jazzy, snazzy
        water dancer.
        I'll be blade
        slicing great arcs
        flashing silver.
        Bird—swift
        swooping through blue . . .
        please let him help me. Now.

I hear her snuffling,
shouting, her hand
grabs at my wrist.
Panic. I want to push her,
get away. But say
'you fancy me then, darling?'
winking at the others.
We laugh; the moment passes.
Don't suppose she noticed.

GIRLS AWAKE, ASLEEP

Young girls up all hours
devouring time-is-money on the phone:
conspiracies of mirth,
sharp analyses of friends' defects,
confession, slander, speculation
—all the little mundane bravenesses
that press the boundaries
of what can be thought, felt and talked about.
Their clear-voiced punctuation rings
up stair-wells, to where parents toss
and groan, a sense of their own tolerance
some consolation for short nights, long bills.

Young girls in bed all hours
fathom sleep oceans,
drink oblivion with their deep breaths,
suck it like milk.
Curled round their own warmth,
they fat-cat on the cream of sleep
lapping dreams.
For this, they will resist all calling.
Surfing the crests of feather billows
they ride some sleek dream animal,
pulling the silk strands of his mane,
urging him on.

PICCADILLY LINE

Girls, dressed for dancing,
board the tube at Earls Court,
flutter, settle.
Chattering, excited by a vision
of glitter, their fragile bodies
carry invisible antennae,
missing nothing.
Faces velvet with bright camouflage,
they're unsung stars—so young
it's thrilling just to be away from home.

One shrieks, points, springs away.
She's seen a moth
caught up in the blonde strands
of her companion's hair,
a moth, marked
with all the shadow colours of blonde.
The friend's not scared;
gently, she shakes her head,
tumbles it, dead,
into her hands.

At Piccadilly Circus they take flight,
skimming the escalator,
brushing past the collector,
up to the lure of light.

CHANGING THE SUBJECT

1 *The Word*

It started with my grandmother
who, fading unspeakably,
lay in the blue room; disappeared
leaving a cardboard box,
coils of chalky-brown rubber tube.

I inherited her room, her key.
The walls were papered bright
but the unsayable word
seeped through; some nights
I heard it in the dripping of the tap.

I saw it in my parents' mouths,
how it twisted lips for whispers
before they changed the subject.
I saw it through fingers
screening me from news.

The word has rooted in my head
casting blue shadows.
It has put on flesh,
spawned strong and crazy children
who wake, reach out their claws.

2 *Out-Patients*

Women stripped to the waist,
wrapped in blue,
we are a uniform edition
waiting to be read.

These plain covers suit us:
we're inexplicit,
it's not our style to advertise
our fearful narratives.

My turn. He reads my breasts
like braille, finding the lump
I knew was there. This is
the episode I could see coming

—although he's reassuring,
doesn't think it's sinister
but just to be quite clear . . .
He's taking over,

he'll be the writer now,
the plot-master,
and I must wait
to read my next instalment.

3 *Diagnosis*

He was good at telling,
gentle, but direct;
he stayed with me
while I recovered breath,
started to collect

stumbling questions. He said
cancer with a small c
—the raw stuff of routine—
yet his manner showed
he knew it couldn't be ordinary for me.

Walking down the road
I shivered like a gong
that's just been struck
—mutilation ... what have I done ...
my child ... how long ...

—and noticed how
the vast possible array
of individual speech
is whittled by bad news
to what all frightened people say.

That night, the freak storm.
I listened to trees fall,
stout fences crack,
felt the house shudder as the wind
howled the truest cliché of them all.

4 *In-Patient*

I have inherited another woman's flowers.
She's left no after-scent, fallen hairs,
no echoes of her voice,
no sign of who or how she was

or through which door she made her exit.
Only these bouquets—carnations,
tiger lilies, hothouse roses,
meretricious everlasting flowers.

By day, they form the set in which I play
the patient—one of a long line
of actresses who've played the part
on this small white stage.

It's a script rich in alternatives.
Each reading reveals something new,
so I perform variously—not falsehoods,
just the interpretations I can manage.

At night, the flowers are oracles.
Sometimes they seem to promise a long run;
then frighten me with their bowing heads,
their hint of swan-songs.

5 *Woman in Pink*

The big, beautiful copper-haired
woman in the next bed
is drowning in pink.

She wears pink frills,
pink fluffy cardigan and slippers.
Her 'get well' cards carry pink messages.

Her husband brings pink tissues,
a pink china kitten; he pats her head.
She speaks in a pink powder voice.

Yet she is big and beautiful and coppery.
At night, she cries bitterly,
coughs and coughs from her broad chest.

They've done all they can.
She's taking home bottles of morphine syrup,
its colour indeterminate.

6 *Pauline*

Six years ago, refused surgery
—liked her breasts,
wasn't going to be cut up.

Three years ago,
her liver affected,
the doctor talked of a few months,

but she wasn't going to die
with her son away in Africa
—not for two years at least.

She swam a lot, played bridge,
visited the sick,
went to car-maintenance classes.

Last month, a funny turn
—it's reached her brain.
She'd worried she was going mad!

So here she is, in the bed opposite,
going for radiotherapy,
losing her hair.

She sits upright,
excited, laughing,
choosing a wig from a big box.

Multitudes of flowers, cards,
friends of different ages, sexes,
colours, crowd around.

Her husband, who always hopes
to have her to himself,
sulks behind *The Times*,

but she knows best.
Friends are security forces,
and she is in command.

7 *How Are You?*

When he asked me that
what if I'd said,
rather than 'very well',
'dreadful—full of dread'?

Since I have known this
language has cracked,
meanings have re-arranged;
dream, risk and fact

changed places. Tenses tip,
word-roots are suddenly
important, some grip
on the slippery.

We're on thin linguistic ice
lifelong, but I see through;
I read the sentence
we all are subject to

in the stopped mouths of those
who once were 'I',
full-fleshed, confident
using the verb 'to die'

of plants and pets and parents
until the immense
contingency of things
deleted sense.

They are his future
as well as mine,
but I won't make him look.
I say, 'I'm fine'.

8 *Anna*

Visiting time. Anna rises from her bed,
walks down the ward, slowly,
treading glass. She wears
her hand-sewn patchwork dressing-gown,
cut full, concealing her swollen abdomen.

She smiles at people she passes;
pulls her shoulders back,
making a joke about deportment;
waves a skeletal hand
at Mrs Shah, who speaks no English.

Her little girls sit by her bed
in their school uniforms. Too good,
they're silent as they watch her,
tall in her brave vestment
of patterned tesserae,

that once were other garments
—as she was: a patchwork mother
made of innumerable creative acts
which they'll inherit with her robe
and make of them something new.

She stops. We hold our breath.
Gaining time, she whispers to a nurse
then turns, walks back to her children,
smiling. Look, she is telling them,
I'm still familiar. I belong to you.

9 Knowing Our Place

Class is irrelevant in here.
We're part of a new scale
—mobility is all one way
and the least respected
are envied most.

First, the benigns,
in for a night or two,
nervous, but unappalled;
foolishly glad their bodies
don't behave like *that*.

Then the exploratories;
can't wait to know, but have to.
Greedy for signs, they swing
from misery to confidence,
or just endure.

The primaries are in
for surgery—what kind? What then?
Shocked, tearful perhaps;
things happening too fast.
Still can't believe it, really.

The reconstructions are survivors,
experienced, detached.
They're bent on being almost normal;
don't want to think
of other possibilities.

Secondaries (treatment)
are often angry—with doctors, fate . . .
—or blame themselves.
They want to tell their stories,
not to feel so alone.

Secondaries (palliative)
are admitted swathed in pain.
They become gentle, grateful,
they've learned to live
one day at a time.

Terminals are royalty,
beyond the rest of us.
They lie in side-rooms
flanked by exhausted relatives,
sans everything.

We learn the social map
fast. Beneath the ordinary chat,
jokes, kindnesses, we're scavengers,
gnawing at each other's histories
for scraps of hope.

'You did not proper practise',
my cello teacher's sorrowful
mid-European vowels reproached me.
'Many times play through the piece
is not the proper practising
—you must repeat difficult passages
so when you make performance
there is no fear—you know
the music is inside your capacity.'
Her stabbing finger, moist gaze,
sought to plant the lesson in my soul.

I've practised pain for forty years
—all those Chinese burns;
the home-made dynamo we used
to test our tolerance for shocks;
hands wrapped round snowballs;
untreated corns—all pain practice.
Which is fine
so long as I can choose the repertoire.
But what if some day I'm required
to play a great pain concerto?
Will that be inside my capacity?

I've hung the washing out
and turn to see
the door slammed shut
by a capricious wind.

Locked out, face to the glass,
I see myself reflected
in the mirror opposite,
framed, slightly menacing.

No need for wuthering
to feel how it might be
—I have that sepia, far-seeing
look of long-dead people.

Perhaps I wouldn't feel dead,
just confused, lost track of time;
could it be years since I turned
with that mouthful of pegs?

And might I now beat on the glass
with jelly fists, my breath
making no cloud in this crisp air,
shout with no sound coming?

Death could seem this accidental
—the play of cells
mad as the freakishness of weather,
the arbitrary shutting out.

Might there be some self left
to look back, register
the shape of the receding house?
And would it feel this cold?

The curtains said:
what do you fear more than anything?
Look at it now.

A white room.
I lie and cannot speak
and cannot move.
I stream with pain from every part.
I cry, scream until the sound chokes me.
Someone at the door looks in,
glances at her watch, moves on.
No one comes. No one
will ever come.

The lamp said:
think of what would be most blissful
—what do you see?

A white room
lined with books; a window
looking out on trees and water;
bright rugs, a couch, a huge table
where I sit, words spinning from my fingers.
No one comes; time is limitless,
alone is perfect.
Someone leaves food at the gate
—bread, fruit, little chocolate birds.

The moon laughed:
there is only one room.
You choose the furniture.

Because a bit of colour is a public service.

Because I am proud of my hands.

Because it will remind me I'm a woman.

Because I will look like a survivor.

Because I can admire them in traffic jams.

Because my daughter will say ugh.

Because my lover will be surprised.

Because it is quicker than dyeing my hair.

Because it is a ten-minute moratorium.

Because it is reversible.

In my fiftieth year,
with my folded chin
that makes my daughter call me Touché Turtle;

in my fiftieth year,
with a brood of half-tamed fears
clinging around my hem,

I sit with my green shiny notebook
and my battered red notebook
and my notebook with the marbled cover,

and I want to feel
revolutions spinning me apart,
re-forming me

—as would be fitting in one's fiftieth year.

Instead, I hum a tune to my own pulse.

Instead, I busy dead flies off the sill
and realign my dictionaries.

Instead, through the window,
I make a sign of solidarity
at swallows, massing along the wires.

*   *   *

PARTNERS

It was always said—she
was the strong one,
the emphasis implying
something not quite natural.

It showed in her head's angle
inherited from a line of officers
khaki-convincing in the gallery
of family photographs.

She always knew her mind.
He never could decide on anything.
After he died, people said
she'd grown to look like him,

—as if his soul, lacking direction,
had managed a short hop
and settled in that softening jaw,
that bewilderment behind the eyes.

VISITING DUNCAN

I'm on a day trip to our shared frontier,
pass tissue-wrapped daffodils, chocolate
across the gap; my greeting warm, but careful,
taking the measure of your foreignness.
How far have you travelled

in your migration to that other country
whose landscape, customs, I can only guess?
It's an America, dividing you from me
by language much like mine, yet skewed,
stripped. I make conversation:

'Nurse says you went to Hastings yesterday
—I wonder if you remember . . .'
'Was it meant to be memorable?' you ask,
not meaning, I think, to be witty
though later I'll laugh, remembering.

You don't talk about the old country,
little of the new—as you did once
when contrast, loss, were everything.
The discourse you've learned here
is that of emptiness.

You examine the dimensions of the void
with all your old precision; picking up
a letter from your son, 'Do I miss him?
What would be the test? Do I wish
he were here? I think not.'

A woman comes into your room. 'I have
nothing', she says. Just that, twice,
and leaves. Later, another: 'I'm so hungry
—can you give me food . . . nothing all day.'
You break her off some chocolate.

'It's an existence'. You leave
the alternative unsaid. Your final exile
has no reference points; in an hour
you won't know I was here. There's only now;
this only kiss; these hands, holding.

FÜR THERESE

I'll tell you why—

You must understand, since he died
he has that special unreality
that greatness gives; as if he's been distilled
into his Ode to Joy, his Grosse Fuge.

I knew a different man—an embarrassment
to good society, gauche beyond belief!
Tone-poet he may have been, but blind
to the colour of reproof in Father's eye
—he would hold forth about that Bonaparte.
Almost a child—he'd know he'd caused offence
but not know how, nor how to put it right.

And yet, he altered when we were alone.
He loved me, and I felt for once
powerful—great temptation to a girl
expected to transform from father's daughter
into husband's wife. But then I saw
how we'd grow disappointed with each other
—he with my limited capacity
to understand his art; I with his constant
absence in a world that shut me out.

Frau van Beethoven! Now I'm facing death
I sometimes whisper to myself that title,
missed, and wonder if I could have learned
enough to follow him . . . but after all
our names were not intended to be linked
—even the little piece he wrote for me
that might have been my mark on history,
his publisher misread as *Für Elise*.

NIGHT HARVEST

for Martin

We dredge these small fry
from our separate pools of sleep,
spread them before each other

and sort them, puzzling,
smiling to discover
our several selves in them.

Under water their colours
were subtly different.
Some slipped back as we lifted them

but these are enough, prismatic,
splitting the past, the future
into bright fragments.

We can afford to be extravagant,
throw back the catch,
know it will multiply.

THE BED

When they were young, and she a captive
in her parents' house, he'd climb
in through her window. They'd whisper,
touch, slip together in her narrow bed
until the rooster pulled them separate
and sent him, singing, to the field.

Their marriage bed was ample.
Child after child was born in it
until, pushed to the side, he played
the hero in the field, the tavern.
Resentments multiplied; the bed was
for sleeping, back towards back, alone.

In middle age, as she gained flesh,
he lightened, rolled towards the centre.
One night, he floated from his dream,
found her arm curved around him,
hand tucked under his side,
and she was murmuring an old song.

In their seventies, he took a saw,
halved the bed's width. Climbing in
earlier each day, they found
a dozen different ways of fitting,
fusing at last into a shape so perfect
they felt no further need to move apart.

RUBY WEDDING

Forty years this month
since you hurtled round the corner
into me, taking my breath away.

Eye-watering you were
like lemons after long thirst,
a burst of bubbles.

I'd learned to patch my emptiness
with tidy habits,
was comforted by order,

but you—a bouquet of astonishments
a chaos I fought
then learned to mingle with.

Sometimes I'd watch you sleeping,
switchback your breath—even then
you seemed so vivid.

I'd rub your hands
skin turning to plastic, paper,
then to ash.

As I've cleared
your squirrelled papers, ornaments,
order has ticked back into every room.

I have been slow
to cast off from the bed
in which we joined and parted

but now I'm drifting out.
You have breathed my last breath.
My heart is jumping for the two of us.

SEX OBJECT

The Romanies are gone.
No longer eyesore, blot, threat, health hazard,
they're off, leaving a scatter of plastic,
heaps of rusty parts,
the smell of woodsmoke.

But in the small hours,
one steals into Jenny Wilson's bed
and lifts her nightdress.
Oh, she murmurs, as his grimy hand
strokes her thigh, testing,
as though he'd rubbed it down with wire wool.
She gleams in the moonlight
ready for further renovation.

His brush is silken. It drips
with fire, smells of molten chocolate.
Flame flows into every hole and crack.
How did I not know this before, she says.

And now, the polishing,
soft, sweet-smelling wax.
He massages her shoulders, hips,
with circular caresses.
She is a race horse,
a Chinese cabinet,
a centuries-old newel post.
Her skin is milk, and steel.

And now—but now it's daylight.
Eyes open on her ordinary room.
She's stretching, thinking—these days
ideological unsoundness
is the most delicious sin there is.

TALLULAH AND THE TRAMP

Indifferent to food,
for three days and sleepless nights
he has laid siege to her.
His fear of indoors gives off vapours
as he follows
into the upper reaches of the house.
He huddles, dour, shabby,
waiting for a sign.

Mistress of contradiction,
she contemplates a hibernating spider
high on a picture rail, yawns
as though finding sleep infectious.
Then rolling, limbs stretched,
begins a slow,
squirmishing
invitation.
He draws near; she claws at him,
spits poison. Dumb,
he resumes his stubborn watch.

The moves repeat themselves
until, by force, or guile,
or her own sharpening need,
he gains her.
His part is played. He runs,
savours cool air, while she
walks by herself again.

WHY I LIE IN THIS PLACE

We were close once.
I knew him better than I knew myself
—the way his lips tensed when he was moved
his knack with children,
the smell of his sweat mingled with the horses'
after a fast ride together.

I made no secret of it.
I heard the envious sniggering,
but I sought him out, and he, me,
I swear it. We drank and whored together.
We discussed court business,
talked sport and strategy

—until he changed. In such little ways
at first—mere moments of distraction,
the smile a shade less warm—I didn't think.
And then, as I was talking once, I caught
a glance, thrown by him to his equerry,
as if to say 'that's typical, you see'

and the world somersaulted—suddenly
no longer partner, fellow-witness,
but object, irreversibly split off.
I left the city then, grew hard.
Indifferent to death, I flourished
in far-flung campaigns.

The people sought me out, asked me
to lead the revolt against him. I'd heard
he had become cruel, a voluptuary, oppressive
—they had real grievances. And yet
their quarrel wasn't mine. Had it not been
for that flick of the eyelid

—the hinge on which hung
love and hate, peace and war, the fate
of princedoms and ten thousand little lives—
I'd have bent my strength to his;
what followed would have happened differently.
As it was, I used the masses' anger.

Or was it the reverse?
Perhaps the swell of history
would have rolled on without my part in it.
I only know, but for that look,
these many hundred years,
I should have been lying next to him.

THE CHAIRMAN'S BIRTHDAY

The day before, my father
had visited the butcher's shop himself
to choose the calf's head.

Our pastry cook was gone
(I hadn't thought of him as an Enemy)
so the dessert course would be ordinary.
But the calf's head! Father's speciality.

He held it up to admire its whiteness;
I shunned its eyes, its open baby mouth.
He plunged it in the bubbling pan
covered closely with a piece of cloth
to stop it turning black.

Meanwhile the sauce—Madeira demi-glace—
quenelles of minced truffles,
sautéed cocks' combs . . .

and when the head was lifted steaming,
placed on a platter, he surrounded it
with mushrooms, halves of hard-boiled egg,
sweetbreads, its own sliced tongue and brains,
poured on the sauce, carried it, glistening,
ceremoniously to table.

*

Because they took him two days later
that evening runs on a continuous reel
inside my head; a melodrama lit by Eisenstein.

Under the chandeliers
Father walks, bearing the dish,
to comic tuba music—though then
it seemed triumphal, dignified.

Cut to the Chairman, who shifts his eyes
as he thanks Father for the feast
while quavering violins,
my father's sweating, deferential forehead,
seem to interrogate the future
and find their own reply.

And then the epilogue
—because I've wondered ever since
if he had slaughtered me,
served me with miraculous garnishes,
would he be living now?

FATHER CHRISTMAS

He'll have left his reindeer
on the roof
—they wouldn't like to struggle
with the chimney.

By now, he'll be
drinking the sherry,
the pie scattering crumbs.
He'll cross the living-room—

at last, the groan
of stairs, handle's squeak.
Amazing he knows
which room is hers.

Eyes tight, she lies
stiff, listens; shuffle,
stumble, breathing hard
from his long journey.

She peeps; she knows.
He hushes her.
And now his weight,
a sort of fumbling.

She's whimpering;
it's her fault
for opening her eyes
before morning.

ZANOOBIA*

I am the scapegoat, the sacrifice
stuffed with the fruits of sin.
There is no expiation, anywhere.

I am the black sheep
the family don't recognize as theirs,
and hope to lose in Africa.

I am a whore,
tricked out with a sassy southern name;
men pump their self-hatred into me.

I'm Typhoid Mary on a monstrous scale.
Steeped in sick substances,
I threaten death to thousands.

I am a vagrant, traipsing the world,
hugging my wretched load
as though it were precious.

For me there is no happy home-coming.
Welcome nowhere, I'm forced to turn,
drag myself away.

*The cargo ship, Zanoobia, carrying 11,000 tons of toxic waste, was
refused entry to several countries during 1988.

NEWS ITEM

I'm here in Noi Serpo to view
startling developments designed
to eliminate dependency
by 1992.

        And now
I'm in a civic hospital
where patients pay for treatment
by repacking used syringes, patching
sheets, and other tasks.
Here is someone writing labels
in her oxygen mask;
while down the hall I'm told these men
will be fit for discharge when
they've done their course of O T,
making tables . . .

        Now we've come
to this old people's home, a hive
humming with enterprise! I gather
twenty-five little businesses
operate here. The profits go
to run the home — the most successful
get baths, and three cooked meals a day.
Even the senile pay their way . . .

        And it's freezing
on this piece of land, where men,
handicapped I understand, are out
on work experience. Strung together
with leather straps for their own safety,
they're helping build a conference suite.
That faint shout you can hear
is a work-song, 'koti feeri,
koti meeri', which I'm told
means 'We dig, we eat' . . .

Apparently
the mental patients proved a problem
until someone devised a plan
to make them even crazier! Now,
as you can see, instead of
sitting on their beds, catatonic,
we have a lunatic fantasia
—hallucinogens have turned them
into first-class entertainment.
The public queues—especially now
zoos have become uneconomic.

Before returning you to London
—you may be asking, what about
children? Well, I asked the same,
as I hadn't seen a child
anywhere. They told me
it's against the rules to visit
schools. Only once, I heard
a child's voice in the distance.
'Weep, weep', it went.
But when I asked,
no one knew what the word meant.

CHRISTMAS CIRCULARS

This is the season when the myth-makers
play Holy Families—their filtered lives
appropriately merging with the stream
of set Nativities, Madonnas, doves . . .

'Robert has been promoted yet again!
We're all extremely proud of him, although
it means he has to travel quite a lot.
Sam's football-mad, but passed his Grade 5 oboe.

Jean took an evening class, Renaissance art
—meals in the oven, but we were amazed
at all she knew on our super stay in Rome.
Beth triumphed in GCSE—six As.'

And from the emigrés: 'We came in June
. . . appalled at how run-down England's become;
no really open space . . . how did we stand
the weather all those years we lived in Brum?

We have a lovely place near Armidale.
Vanessa's tennis champion of her school.
You wouldn't know us, we're so brown—think of us
all celebrating Christmas round the pool!'

They say, between the lines where they regret
there isn't time to write to each of us:
Our life is an accomplishment, a pearl
whose perfect shape and sheen deserve applause.

It's hard, of course. But when we see our lives
reflected here, we're almost led to think
that that's reality. So though poor Jean's
on Valium for her nerves, and Robert drinks;

and though the children quarrel constantly
and Beth won't smile, and sometimes wets the bed;
and though we often seem to feel the draught
knife through well-fitting doors—it can't be said.

56

AT THE FEAST

We sipped watercress soup,
relished the gigot of lamb
with mustard sauce,
crunched cool salad
—endives, bean shoots, almonds—
savoured the crème brulée

and all the time, we talked,
eight of us round an oval table.
While the flames of candles
flattered our skins
we rinsed the ills of the world
around our claret glasses.

Then I heard rough feet
snagging in the pile of the carpet
tss...tss...like hurt breath;
the voices, a stubborn mantra
as if one word must do for everything,
and that long drained of sense.

They were coming down the hall,
they poured into the room, shadows
stalking the walls, hands stretched.
I almost thought it was a dance,
the flickering of the candles
made them move so.

Plates clashed and silenced them;
the walls were calm,
my friends were quite themselves.
But my wine seemed opaque,
and curled among the grapes and peaches,
a baby, a shrivelled fig.

MOVING TARGETS

Yesterday, as usual,
Tom was in his meadow pitching hay,
movements spare as the fork
that's done him and his brother sixty years.

In the next field, the big farmer's plane
black against sun, crop-spraying.
At the sound, rabbits everywhere,
crazy, zig-zagging . . .

I remembered people in a news-clip
running, burning as a plane
wheeled away; the terror
in their broken trajectories.

Gusting wind harried the cloud
wide across the land, dispersing
as the old man worked on,
tracing his effortless parabolas.

He's not in the field today.
Many villagers are ill
—throats, eyes, faces burning.
I ask if they'll protest. They hesitate,

'. . . wouldn't want to cause bad feeling
—and real harm wouldn't be allowed.
It's not the first time.
The burning doesn't last.'

I look at their soft faces,
their sensible shoes; imagine them
—Joan, the Andersons . . . and Tom—
in a news-clip, running, burning.

INTER-CITY

for Anne Harvey

Opposite me
a fat brown man
is crying
fat glass tears
on to his Fair Isle pullover.

Needles of rain
mean-streak the landscape.
Bricks ... allotments ... bricks ...
and the fat brown man
sits opposite me, crying.

Perhaps he thinks
no one will notice
if he keeps his eyes closed,
his face forced
into composure.

And everyone is
not noticing,
minding our business,
our English tact
unbreachable.

Or is it cowardice,
fear of rebuff?
And is it worse
to string the tears
into a narrative?

GHOST STATIONS

We are the inheritors. We hide here
at the roots of the perverted city
waiting, practising the Pure Way.
Listening to ourselves, each other,
we find the old soiled words won't do;
often we can only dance our meanings.

Deep in the arteries of London, life
is possible—in the forgotten stations:
York Road, St Mary's, Seething Lane...
I love the names. Each day, we sing them
like a psalm, a celebration
—Down Street, British Museum, City Road.

We live on waste. After the current's off
we run along tunnels, through sleeping trains,
ahead of the night cleaners. We find chips,
apple cores (the most nutritious part),
dregs of Coke. On good days, we pick up
coins that fit the chocolate machines.

Once I found a whole bag of shopping.
That night we had an iceberg lettuce,
a honeydew melon, tasting of laughter.
And once, an arbutilon—its orange
bee-flowers gladdened us for weeks.
Such things are dangerous;

then, to remind ourselves, we read
the newspapers we use as mattresses.
Or gather on the platforms,
witness the trains as they rip past
(our eyes have grown used to the speed).
Almost every known depravity

is acted out on trains—rape, drunkenness,
robbery, fighting, harassment, abuse.
And the subtler forms—intellectual bullying,
contempt, all the varieties of indifference...
We've learned to read the faces;
we need to see these things, simply.

The travellers only see their own reflections.
But lately, a few in such despair
they cup their faces to the glass, weeping,
have seen the ghost stations
and though we're always out of sight,
they sense our difference and find their way.

Our numbers are growing, though there are
reverses. Some lose heart, want to leave.
We can't let them—we keep them all
at Brompton Road, carefully guarded,
plotting uselessly, swapping fantasies,
raving of sunlight, mountains or the sea.

One day, we'll climb out, convert the city!
The trains are full of terrible energy;
we only have example, words. But there is
our chant to strengthen us, our hope-names:
Uxbridge Road, King William Street,
South Kentish Town, South Acton, Bull and Bush . . .

OXFORD POETS

Fleur Adcock

James Berry

Edward Kamau Brathwaite

Joseph Brodsky

Michael Donaghy

D.J. Enright

Roy Fisher

David Gascoyne

David Harsent

Anthony Hecht

Zbigniew Herbert

Thomas Kinsella

Brad Leithauser

Derek Mahon

Medbh McGuckian

James Merrill

Peter Porter

Craig Raine

Christopher Reid

Stephen Romer

Carole Satyamurti

Peter Scupham

Penelope Shuttle

Louis Simpson

Anne Stevenson

George Szirtes

Grete Tartler

Charles Tomlinson

Chris Wallace-Crabbe

Hugo Williams

also

Basil Bunting

W.H. Davies

Keith Douglas

Ivor Gurney

Edward Thomas

63